How To Gather and Use Data For Business Analysis

M.L. HUMPHREY

SELECT TITLES BY M.L. HUMPHREY

CONTENTS

INTRODUCTION

There's a lot of buzz these days around big data, data science, and data analysis. And for good reason. Because the more you understand about your business and your customers the more profitable your business can be.

For example, say you have two products, widgets and whatchamacallits, but you don't track the product-level revenue or the product-level expenses. All you know is that last quarter you made $10,000 and you spent $8,000 and that you sold a lot more widgets than whatchamacallits.

You might be tempted given that limited amount of information to think that you should abandon whatchamacallits and sell only widgets. But if you had detailed information about each product you would instead see that you made $6,000 selling whatchamacallits and $4,000 selling widgets and that your expense for selling those whatchamacallits was only $1,000 while your expense for selling widgets was $7,000.

Put those product-level numbers together and you'd see that you made a profit of $5,000 on your whatchamacallit sales and lost $3,000 on your widget sales.

Given those numbers, you'd want to abandon widget sales and focus on whatchamacallits. Or you'd want to fix the issue with why widgets are costing you more than they're making you.

But do you see there why having good information about your business can be so valuable?

Often we make decisions from the gut—"I sure sell a lot of widgets, I must be good at that"—instead of using factual information that maybe tells a different story. That's why collecting good data and using the data you have to the best effect is so important and crucial to success. That's why all the buzz.

But data science is a complicated field. There's a lot of that goes into doing it well and not everyone has the type of mind that can process data at those levels. I myself have dabbled in learning R programming and taken a few courses on big data and data science and while I found it interesting and challenging I realized that it wasn't really what I needed as a small business owner. I'm not trying to parse social trends by mining

Twitter feeds, for example.

I just want to know what product sells the best so I can make more of it.

So what I'm going to cover for you in this book are some very basic principles around how to collect information, how to store that information, and how to use that information effectively. These principles are a good foundation for even highly complex analysis, but you don't need to do that highly complex analysis to get a competitive edge. You just have to collect good information on your business and do so in a way that lets you use that information to better understand what you're doing.

At the end of this book I'm also going to touch on a concept I fell in love with in my MBA program called activity-based costing. It won't be a deep dive—I'm not trying to turn you into an accountant—but it'll be enough of a discussion of the concept for you to understand how collecting the right data about your business and using it properly can transform your business. (Or so I hope.)

The information in this book is based upon my experiences as a regulatory consultant to Fortune 500 financial institutions and, prior to that, as a securities regulator (the person who told those people what rules they were violating).

A lot of the work I did involved taking data

about a company's operations and figuring out what that meant in terms of what was happening at that company. My role was more risk-based than financial, but the principles are the same. (So, for example, I was more interested in if they had a registered representative who was stealing from clients than I was in whether they were going to improve their earnings that year.)

But whatever your focus, it's still all about data at the end of the day.

The guidance in this book comes from real-life experience and frustration. I have made half of the mistakes I'm going to tell you about in this book and had to deal with the consequences when someone made the other half. Not fun. I hope that by reading this book I'll spare you some of that pain.

Alright, then. Now that you know what we're going to cover in this book, let's dive right in.

UNDERSTAND YOUR BUSINESS

In order to effectively apply the principles in this book, you need to understand the nature of your business or you need someone on your team who does.

The types of questions you need to be able to answer include: Who are your customers? Where are they located? What do you sell? Where do you sell that product? How are you structured? Where do you collect information and who do you collect it from? How is that information collected? How is that information stored? What other systems do you need to interface with? How are you going to use the information you collect? Do you need to use that information for financial purposes? For reporting purposes? What requirements are you under with respect to your information? Is any of it confidential?

Before you can work with your data and figure out the best way to collect it, store it, and use it, you need to have a grasp on the business in question. Working with data in a vacuum is a very, very bad idea. It leads to mistakes and frustration from your stakeholders (clients, vendors, internal staff, etc.) It also means that you won't know the information you need to collect, you won't know how to store it so that it can be used most effectively, and you won't be able to judge the results of any analysis you do.

Let me give you a few examples of why it's important to understand your business.

Example 1:

I'll start with my publishing business. It's small enough that I can use Excel workbooks and an Access database to track my revenue and expenses. I'm also the only one who inputs the data, analyzes it, and uses the results. I have data that comes in from multiple sources, but once that data is input into my Access database, I'm the only one that works with it.

I can't control the data I'm provided by the companies I use to sell my books. Amazon isn't asking my opinion about what I'd like to see in my reports. They give what they give and I have to work with that. If I were approaching things from a pure data analysis perspective and not a business perspective, I might just stop there.

And, sure, I could link books by title across the various companies where I sell my books and see total units sold and amount earned, but that wouldn't be enough to effectively run my business.

Because if Title A earned me $5,000 and so did Title B, but Title A took me twice as long to write and had advertising expenses that were twice as much as Title B, those titles are not equal. I should focus on Title B not Title A in that situation.

But the companies that sell my product can't tell me that. I need to track how long it takes me to write and edit every title. That lets me calculate a per hour revenue rate for each title.

I also need to track advertising spend on a ti-tle-level basis. That lets me calculate a per hour profit and loss for each title.

Both of those measures are essential to my making sound business decisions.

But I wouldn't be collecting that information if I didn't know my business well enough to understand that it matters how long something takes for me to produce and that it also matters how much I have to spend advertising each product in order to sell it.

Another example from my publishing business is in how I aggregate titles at the series level.

Some titles sell fine on their own. They're standalones and can be viewed without looking at any other title.

But other titles are related and should be considered part of a series. I have a three-book fantasy series that I've published. I only advertise Book 1, but that advertising on Book 1 leads to sales of Books 2 and 3. So I can't just look at profit and loss on a per-title basis in that situation. I have to look at profit and loss across all three titles combined.

Knowing which titles are standalones versus which titles are series-based and knowing what ad strategy has been used for a particular series is all business-level knowledge. Someone really good at data analysis *might* be able to parse that relationship out but the business-side person *knows* it.

In this case I'm both people. I'm the business person and the data person. But that's rarely the case for a larger company.

So if you are in a larger corporate setting and you're trying to build a system that tracks data or uses your existing data, it's essential that you involve someone with business knowledge in addition to your data people.

Let me give you another example.

Example 2:

At one point in time I was tasked with leading a team that looked at available data to identify those individuals who were the highest risk for harming customers.

Information on these individuals was stored in a number of different systems and in a number of different formats. There was also information we didn't have on them that a later stage of that project considered acquiring. (For example, credit scores.)

The data people could have just taken the information in the systems and worked blind to find patterns. For example, they could have taken people who were "terminated for cause" (it's a specific defined term where I come from) and then looked back to see if there was something about those individuals that differentiated them from others.

But our tech people honestly didn't even know enough about the process to know that a TC was something to look at. That was business-side knowledge. They also didn't know that there was overlap between two of the complaint reporting systems. Or that one of the complaint reporting systems carried more weight than the other because of the criteria they used.

And they didn't know that some of the reporting codes were considered administrative versus sales practice. (The difference between "hey, I think this guy stole some money" which is a sales practice issue and "hey, I didn't expect to pay an annual account fee" which is administrative.)

By combining the data guys and their analysis knowledge with the business-side team and their knowledge of how things worked, we were able to achieve the best possible result. The business-side team wouldn't have known where to find that data or how to process that many records even if they could, and they certainly wouldn't have known how to properly apply a regression model to the data. But the data guys didn't know enough about the data to know what to include in the model and what to exclude.

You need both sides.

One last example.

Example 3:

A few years ago I worked on a consulting project for a large multinational financial corporation. We were trying to standardize their data so that all systems across all geographies would be on the same page and be able to interface without any issues. So, for example, we didn't want one system that said "Great Britain", another that said "United Kingdom", another that said "GB", and another that said "UK". We wanted all systems to use one single way of identifying the United Kingdom.

(Or, failing that, we wanted a mapping between the systems that said Great Britain=United Kingdom=GB=UK.)

There are standards that exist out there for things like this. Groups that have gotten together to establish common ways of referring to things and that work together if something changes. One of the main standards is the ISO standard, for example.

And it seems easy enough to say, okay, we'll adopt the ISO standard wherever possible. It's a large international body that sets standards. Sounds good.

But it turns out that some decisions are actually political and vary by geography.

For example, is it Myanmar or is it Burma?

Which you use will depend on your business, your client base, and where that information is acquired.

If you have customers located in that country who fill out a customer information form online, then you need to take into account the way that citizens of that country identify themselves and what the reaction of your client base will be to the decision you make.

On the other hand, if you're mailing information from the United States to that country then you need to use the country name that will get that mail through the United States postal system and through the postal system for that other country.

It's not just a tech decision. It's a business decision. You need both parties involved.

* * *

Okay?

So. Before you collect, use, or analyze your data, you need to understand the context in which that data was created and is being used. And you need to understand not just the data itself, but the business that uses it.

Now let's dive into those data principles that will let you collect, track, and use the information you need in an effective way.

TRACK EVERY VARIABLE YOU WANT TO ANALYZE IN A USABLE WAY

This is a very basic concept. It has two parts to it.

First, you need to record the information you think you'll need.

So, for example, as I mentioned above, I track how long it takes me to write and edit everything that I publish. That's not something that's easy to recreate after the fact.

If it takes me six weeks to write the first draft of a fantasy novel and another four weeks to edit it, how many hours did I actually spend during that ten week period? (Far less than you'd think looking back on it. It certainly wasn't eight hours a day, seven days a week.)

That's why, in my opinion, it's always better to

track too much information than too little. I'd rather have a piece of data that I don't use than not have a piece of data I realize I need later.

Assuming, of course, that the information you're tracking is easy to track. If it's going to take you hours to obtain one piece of information and that information isn't something you can see a use for, then don't track it.

But if it's something that you can easily and readily record in the moment, do so.

This, of course, takes us back to the notion of understanding your business. To identify information to track and how hard it will be to do so, you need to understand what you do, how you might use that information, and the easiest way to record it.

(Don't even get me started on how difficult some companies make it to track simple information. Usually because they didn't actually check with the people who are going to track that information for them, they just told them what they had to do.)

The second part of this concept is that you need to record the information you track in a usable way. That means not scrawling it on a scrap piece of paper that you then promptly lose.

If you plan to do any sort of data analysis then the information you track needs to end up in a database or spreadsheet. And it needs to be tracked the same way every time (using the same or comparable units and the same inputs). It also needs to be tracked consistently. You can't track for five days and then not track for five months, especially if the data is of a type that can't be recreated later.

Going back to my time tracking example, I happen to record my time spent writing and editing by hand in a day planner. But periodically I go back to the day planner and input the information into an Excel workbook that lets me calculate total time, average time, total word count, and average word count per hour.

So, first principle is that you need to not only understand your business, but actually track the information that will be of use to you and in a format that allows you to *analyze* that information.

SEPARATE THE INFORMATION YOU WANT TO TRACK INTO DISCRETE UNITS

Now that you know what you want to track, you also need to spend some time thinking about exactly how you'll record the information you're going to track.

There are a few principles that tie into this portion of things, but the first one is that you need to separate your information out into discrete, analyzable units.

If you think you might care about a specific aspect of something, make it a separate data point.

For example, in my time tracking I track editing and writing time together. That's problematic if it turns out that I need to be looking at writing time separate from editing time. (I don't because

I tend to write a lot of new material as I edit. But if I did, I'd be in trouble.)

Here's another example.

Let's say that you sell widgets in various colors and sizes. There are green, blue, and orange widgets and they come in small, medium, and large sizes. So you can sell a small blue widget, a small green widget, a small orange widget, a medium blue widget, etc.

Now, you could record each of those items in your system in just that way. Item 123 is small blue widget. Item 456 is small green widget. So on and so forth.

If you were systematic about it, you could still do a lot of analysis even with the information stored that way. If size comes first for all products, you can extract size information easily enough, for example.

But it's far better to have each feature recorded separately. So, in this case, you might have product 123 and then in separate fields or columns record the product type (widget), the product color (blue), and the product size (small).

If you have already separated out any attribute that you want to analyze, then it's incredibly simple to run a report of how blue products sell compared to green products. Or how small products sell compared to large products. And you can still easily enough combine attributes to compare how small blue products sell compared to small green products, for example.

The goal of these data principles is to make it as simple as possible to do any sort of analysis. So it's not that you can't back into the same result if you've chunked your data together, it's just that it takes more effort and there can be errors introduced.

In our example above all it takes is one product to be set up as a green large whatsit to mess up the ability to extract size and color information from a product description.

So always try to separate your data into as many discrete units of information as you can.

ENTER ONCE, USE MANY TIMES

You may be looking at that last principle and asking, "But won't that be messy? Won't I end up with all this duplicate information that I have to re-enter everywhere and risk entering it wrong somewhere?"

Not if you build your data tables correctly.

What you should attempt to do at all times is build your data tables so that each piece of information is only entered once. For anyone who has ever used a commercial accounting system or a customer tracking database, you've probably seen this in action already and didn't think much about it.

Let me give you an example using customer account information.

Most customer billing systems will have a screen where you input new customer information. You provide the customer name, address, contact information, etc. in that one location. The system then assigns a customer identifier. Joe Smith becomes account JS123.

If a system is well-built it will then have you look up your customer information from a dropdown menu or search field anytime you want to use that customer information. The system will then go to that original database of customer information and pull the customer-specific information that's needed.

So if you're going to send an invoice to customer Joe Smith you go to the invoicing screen, look him up, and the system pulls in his name and address. You do not re-enter that information. If it's a really good system you can also look up each product you're selling him and pull that in as well from the product database.

The goal with any sort of database or data tracking system should be to record information once and then link to where that information is recorded when you need to use it rather than introduce user error and inconsistencies by requiring users to re-enter the information.

Now, I should add a little note here. There are times when building your system to work that way can be more effort than it's worth. (I am less stringent about this principle for my publishing efforts, for example.) But in those cases at least try to build in some sort of process where you double-check your information to confirm that it was entered accurately.

BE CONSISTENT ACROSS DATA TABLES OR DATA SOURCES

Another important principle when inputting information is to create consistency across users and systems. This applies most to larger corporations that have multiple divisions or multiple companies under one umbrella.

I have worked with entities that had literally thousands of separate systems collecting information across hundreds of entities. A lot of times these systems pre-dated a merger. So Subsidiary A's system was built when it was owned by Company B and therefore doesn't collect information in a way that's compatible with its new owner, Company C's, systems.

Let's use income as an example.

Let's say Subsidiary A collects customer income

and gives each customer the choice of four values: $0 to $25,000, $25,001 to $50,000, $50,001 to $100,000, and $100,000+

But the other companies owned by Company C collect customer income using a choice of six income values: $0 to $20,000, $20,001 to $40,000, $40,001 to $60,000, $60,001 to $80,000, $80,001 to $100,000, and $100,000+

Both options collect information on customer income within the range of $100,000. Anyone over that range is in the $100,000 plus category. Both are effective. One is not right. One is not wrong. They both do the same thing, but in slightly different ways.

The problem comes when you have to compile results across companies or units that use the two different options. You'd be unable to do so without compromising your information because there is no clean overlap between the categories used for each one.

A customer who falls in the $25,001 to $50,000 category under the first option could fall into either the $20,001 to $40,000 category or the $40,001 to $60,000 category under the second option.

After the merger you can change how Subsidiary A collects income information to align it with the other companies, but the old data will stay inconsistent. That will always be an issue you're dealing with. And in mergers that's just life.

But a lot of companies let Subsidiary A set its own parameters and Subsidiary B set its own and so create problems like this that never need to happen.

Let's take this example a step further.

Let's say that all the subsidiaries use the same six income values as in the second example, but that the company decides to launch a new subsidiary for handling high net worth individuals. The minimum income to open an account with the company will be $100,000, although immediate family will also be allowed to open accounts.

The new subsidiary wants to track income but with a cap of $1,000,000 instead of $100,000.

Where this could go wrong is if the new entity were allowed to set parameters of say $0 to $75,000, $75,001 to $150,000, etc. Because those categories create a mismatch between the categories used by the new subsidiary and the existing subsidiaries. Once more we can't cleanly aggregate customer income information across entities.

But there's a way to make it work. One way would be to take the existing categories and just allow this entity to add on additional categories up to $1,000,000 and then map those additional categories to the $100,000+ category when looking at data company-wide. That's the best solution.

But from a business perspective it might not be the most desirable. A unit focused on high net worth individuals may not want ten check boxes for income on their form, five of which don't apply to the majority of their clients. In that case you can compromise and let them choose their own categories but insist that they at least align with the old ones. So they can't use $75,000 as the first category cut off but they can use $40,000, $60,000, $80,000 or $100,000.

That approach may limit how you can roll up your customer data, but at least everything fits together.

* * *

So if you have multiple units or entities and you need to combine their data at some point or expect that you'll want to make comparisons across units or entities then establish corporate-level standards that all units or entities have to fit into.

Within those standards you can allow some flexibility, but ideally you'd still have each unit or entity map to some level of the corporate standard. And if that happened you'd also need to maintain a mapping between the unit- or entity-level categories and the corporate-level categories so that reports or comparisons can still be easily generated across units or entities.

USE CATEGORICAL VARIABLES INSTEAD OF FREE TEXT WHERE POSSIBLE

Now let's talk about a couple tips on how to collect the data you need.

A categorical variable is one that has a limited number of values. So in the example above where customers could pick one of six options for their income level that was categorical. Rather than have the user input whatever they want for income, you have them choose the bucket they belong in.

Why do this?

Because when you allow users to choose their own values it makes it much, much harder to aggregate your results and analyze them. For example, a free text version of asking for income might say: "How much do you earn per year?" and include a line where a user could enter any value they wanted, including text which might return results such as "$50,000", "$50K", "fifty thousand", and "not enough".

All of those could be the same answer, but how do you know that without looking at each one? And that "not enough" answer doesn't actually answer the question, does it?

A categorical version of asking for income might say, "How much do you earn per year? Check the appropriate box" and then include a series of check boxes with value ranges next to them. It then becomes very easy to analyze how many users checked each option.

Now, there is a middle of the road option where users are given the ability to input any response within certain parameters. So you might ask for income and then require that the income provided be a number of up to seven figures with a .00 at the end to indicate that it's currency.

But chances are checkboxes, because of the limited number of choices, will give you the most consistent responses to your question. Choosing the right values for those checkboxes, however, takes us back to the need for business knowledge. If I have a high net worth clientele but my top categorical value choice for income is "over $100,000" that's probably not going to do me much good because almost all of my customers will be checking that box.

So if you're going to use categorical variables, you need to know what choices make the most sense for that business.

There is also the issue of information overload. Be careful just how many choices you give a user. Five or six is probably ideal for most circumstances.

Understand, too, that by using categorical variables you do lose precision. Looking at our income example again, you won't know where in the range a particular individual falls. If they check that their income is $25,001 to $50,000 you don't know whether it's $25,002 or $49,999. But oftentimes losing that precision is far better than dealing with values that range from "not your business" to "$50K" to something illegible or nonsensical.

USE DROPDOWN MENUS WHERE FEASIBLE

In the same way that categorical variables can be useful, I also recommend using dropdown menus where feasible. This is another way of limiting the potential values that users provide. It's essentially using a categorical variable but in a list format. It works better than checkboxes for situations where there's a larger number of potential choices.

Think of all the forms you've filled out online where you had to provide your address, including your state of residence or country of residence. A checkbox in that situation would be unwieldy, but allowing users to type in their own response is equally problematic.

I once worked with a dataset that was created using a free text field for country information. It had hundreds of thousands of values that users had provided. And for individuals who lived in the United States I saw about forty different variations including United States, USA, U.S.A., U.S., US, Uited States, U.SA and all sorts of other very creative often misspelled entries that all meant the same thing but were not.

And, yes, you could scan those values and know with a fair degree of certainty that all of those individuals had meant the United States. But you can't be certain. And running any sort of automated query using that data would be incredibly challenging because you'd have to make sure that all possible variations were covered.

A dropdown menu that lists one approved version of each entry is far more efficient.

Anytime you can standardize the list of values people will enter and turn that into a simple dropdown menu for them to select from instead of allowing free entry, you should.

Now, this isn't always as simple as it seems. Because you have to choose the values that go into that dropdown and there are sometimes situations where the choices are not easy to make. I mentioned above the issue with Burma and Myanmar. You could maybe list both if you wanted. Or you could choose the standardized list that most aligned with your corporate situation.

But there are others you may not even have

thought about. Which salutations should you include in your dropdown? Ms., Mr., Mrs., Dr.? Okay. But what about all the potential military titles out there like Lt. Col. or Lt. Col. (Ret)? What about other titles or salutations that maybe only apply to one individual, like President?

When you start to dig in on these options, it can get very complicated and/or very political very fast. But you have to do it unless you want to spend a lot of time and energy on the back end cleaning up your data.

One more point to make with respect to dropdown menus. The more options you have available the more it makes sense to list the most common options at the very top followed by an alphabetical listing of all choices. Once again, you'll have seen this in online forms you fill out where maybe the United States, United Kingdom, and one or two other countries are listed first. As a user of systems that haven't done this (hello, Bowker), I can say it's much appreciated when it is done that way.

CONSIDER USING AN OTHER OPTION

This next principle is not one I'd always recommend using, but it is one I do think you should consider.

If you use dropdown menus or categorical values that limit a user's inputs to a set number of choices, you should also consider whether it makes sense to include an "Other" option and then the ability for the user to input their own unique value.

Ideally this field would only be used on rare occasions when it turns out that you missed some rare exception or occurrence. So this could be where a member of European royalty inputs their salutation because you decided not to list all possible salutations for European royalty in your dropdown menu of choices.

It can also help to identify situations where a question is poorly worded or users don't feel their preferred response has been included.

This also allows you to catch situations where there has been some sort of change in the available options that your dropdown or checkboxes fail to capture.

So, for example, if you had included Yugoslavia as a country on your country list and failed to update that list when Yugoslavia split into Bosnia and Herzegovina, Croatia, Macedonia, Montenegro, Serbia, and Slovenia. An Other option would let you catch that change when it occurred.

It can also prevent you from dirtying up your data with bad answers.

For example, if you give your users the choice of three income ranges—$0 to $25,000, $25,001 to $50,000, and $50,001 to $75,000—but you have a user who falls outside of those choices and you make choosing one of those three options mandatory, then you are forcing that user to choose an incorrect answer. They'll choose the closest value they can and you'll never know that your list of choices was inadequate.

If you do include an Other option, you need to monitor the responses you get to see why. If it's the rare exception, fine. But if it's poor wording of the question, you need to fix the question.

And if you're missing a legitimate choice, you need to amend your list of choices. It won't retroactively fix your data. You'll be stuck with those responses as they were entered. But at least you can improve the situation going forward.

Of course, you may also find that some users are just crazy. Don't add every single thing someone puts under Other into your dropdown of choices. Evaluate each answer from a business perspective and decide whether it belongs in the list of available choices or not.

One more point here.

If you don't allow room for an Other option and you make an answer mandatory you should carefully monitor what answers you are getting and have someone with knowledge of the business gut-check the results.

I've seen more than one situation where account representatives told customers to just choose X to get through a question even though X wasn't a valid answer for that customer. If you box your users in, they will find creative ways around you.

KEEP YOUR VALUES UP TO DATE

Following on from what we just discussed, if you do use categorical values or dropdown menus, you need to monitor for developments that will impact those values and promptly update your fields as needed.

I gave the example above about Yugoslavia. Another one I can think of is when the USSR dissolved. Countries change their name. Countries merge. Countries dissolve. You have to be prepared to account for that.

And your business may change. What made sense for potential ranges to use for a categorical variable two years ago may not make sense now. You may expand your business lines. You may contract them.

Another example I can think of is tax brackets. In the United States there were recent changes to the tax laws which changed the tax brackets for individuals. Any entity that was asking its customers for their tax bracket using checkboxes would need to promptly update its forms or risk having customers not know how to answer the question anymore.

So once you commit to using dropdown menus or categorical values, you also need to commit to creating a process for monitoring for changes in those values and updating them in a timely manner.

The bigger the company, the longer the lead time on something like that. You can't wait until a customer complains to ask your tech team to please add a new country to the new customer system and expect them to do so within twenty-four hours. That's not how it works.

And failing to update your information in a timely manner will lead to having junk data in your systems which impacts your ability to properly perform analysis.

This is not a set it and forget it decision. It's an iterative process.

TAKE CARE WHEN MAKING CHANGES TO YOUR DATA

Having just said that you need to monitor your systems and update your categorical values and your dropdown menus in a timely manner, let me also say that you need to take care when making changes to the way you collect or store your data.

Chances are, at some point in time someone in your company will be tempted to "fix" your data in some way. Sometimes this needs to happen, such as the situations where countries cease to exist and new ones come into existence. Or where there's a shift in your business or you discover that a question you're asking isn't collecting the information you want.

But sometimes it's just someone feeling they can make the world better by changing a field name, for example.

If the change is not essential (and renaming a field often is not), seriously reconsider making the change. Because if that data table or data is used anywhere else, you could be creating significant problems for other people with your simple change.

Let me give you an example.

Amazon loves to do things like this. At least it feels that way.

Every month they provide a report to their publishers that lists all of the sales that publisher has had for that month. And a couple years ago they decided to rename one of the fields in that report to something insanely long and complicated that very precisely defined what was included in that field.

It just so happens that I was using the old field name to split out two types of sales that needed to be handled differently. So they made this change—without notifying any of us because who are we to them—and it broke my database reports. I was suddenly getting error messages.

And because they had done this without telling anyone about it, I had to dig in and figure out what they'd done and why my report was now broken, which cost me a couple hours of effort. Sure, I'd built the reports, but a year or two later I had to walk through each step of that report generation process to find what was no longer working. I was not happy.

At the same time, there was a data vendor who charged for a service that took Amazon's monthly reports and turned them into pretty, shiny graphs. That vendor wasn't made aware of the change either. The first they knew about it was when their users started contacting them and demanding to know why their pretty, shiny graphs weren't working anymore.

I don't know when they started working the problem, but they were down most of that day with users going onto forums and discussing what had happened and threatening to ditch their product.

Hours wasted. All because someone decided to be more precise in the name of a field in an Excel file that was sent to end users.

This same thing can happen when you insert a new column into a report, delete an old column from a report, change the question behind the data you're collecting, change the options available for a specific field, etc.

Always think about where your data is being used and how your changes are going to impact that before you make any change and then ask yourself whether it makes sense to do so. (And, once more, get user and business side input, too. And don't just listen to the loudest voice in the room either. Get a consensus.)

DON'T CHANGE YOUR RAW DATA

Another principle for handling data is that you should never change your raw data. Whatever you get, whenever you get it, in whatever form it comes in, is what you have.

If you need to make adjustments to that data at some later point in time, do not touch the raw data. Make your changes on a copy of the data.

Let me give you an example of where this went wrong.

I don't know for a fact that this was a situation with the raw data since we were provided an extract from a database, but it very well could have been. It felt like it was.

What we were dealing with was a series of responses to a questionnaire. At one point in time the questionnaire had asked "Do you engage in either Activity A or Activity B?" A yes answer to that question meant either A, B, or A and B was true for that entity. You couldn't know from the answer to the question which it was.

At some point in time the question was changed and split into two different questions. "Do you engage in Activity A?" and "Do you engage in Activity B?" Now it was possible to know whether the answer for a given entity was A, B, or A and B.

So far, so good.

The problem was in how the data folks handled the integration of the two questions.

Rather than have the original responses as one entry in the database and the answers to the new questions as two new and separate entries in the database, the data folks used the same field for answers to the question prior to the change and for answers to both questions after the change. Now, this still could have worked if they had recorded that "yes" answer from before as "A, B, or A and B" and then the new "yes" responses were recorded "A", "B", or "A and B". So one answer before becomes three answers after.

But they didn't. What they did is overwrote all prior responses to that first question as "A and B". That was inaccurate. Because a company could have been doing only A and answered yes to that question before the change. Or they could have been doing only B and answered yes. (And if you looked at the data after the change it was very unlikely that any company had been doing B, so this was a material issue.)

The data folks made a mistake. And if they did so on the raw data, there was no easy way to fix that mistake. All answers to that question prior to the change were permanently flawed from that point onward. If you erase an answer and put in a new one, it's not always a simple process to go back to the original.

That's why best practice is to leave your raw data untouched. If this issue had been created by some data transformation outside of the raw data then it would be easy enough to change the rule that was applied to the raw data and run the data extract again.

So, store your data somewhere, extract what you need from the raw data, and then fix anything that needs fixed at that point.

And I'll say this, too.

I don't care if you think you know what someone meant. "Oh, they put U.S. I know they meant the United States." You don't know that. You can assume, and in your analysis you can make that change, but never have the hubris to overwrite what someone actually said. You should always be able to go back to what you were given and then walk someone through to what you ended up with.

DOCUMENT YOUR DATA
CLEAN-UP AND ANALYSIS STEPS

That leads us to our next data principle. It is a best practice to document everything you do to take your data from its raw form to the form you use for your analysis, and to document what steps you go through in performing your analysis.

Someone else should be able to take that exact same raw data and reach the exact same final conclusion that you did.

Do not do what I did once when I thought I was doing a one-off project and adjust your data without documenting the steps you took to clean it.

In that particular situation it became an ongoing report we needed to provide and that first analysis never did match up with the later analysis, because I hadn't documented my steps. (And, quite frankly, couldn't do that level of clean-up every single time they wanted that particular report. We had a training issue with staff inputting their responses that I tried to fix on the back end, something that never goes well.)

When I was in a corporate setting I did a lot of my analysis work using Excel because it's what I knew and what was available to me, but this is where using programs like R can come in really handy. With R you can basically write an entire script that goes and grabs your raw data file, performs all the clean-up and analysis you need, and then exports your final report. Anyone who wants to see what you did can review the script and decide for themselves whether everything was done properly. They can also replicate the analysis with just the click of a button.

It's a far better option than what I used to do in Excel. (Remember, this book is written in part as a result of the mistakes I made when I didn't know better. As someone once said, "most rules are written in blood.")

SAVE VERSIONS

If you are going to work in a program like Excel that doesn't automatically document all of your data clean-up and analysis steps, I'd highly recommend saving versions of your data file as you work your way through the clean-up and analysis steps.

For example, let's say that you need to do calculations A, B, C, D, and E on a set of data.

I would bring in a copy of the raw data and clean up that data with formatting or sorting or whatever else I felt needed to be done first. I'd then save a copy of that file. Next, I'd set up calculation A and save another copy of the file under a new name. And then I'd set up calculation B in addition to calculation A and save a version of that file and so on and so on until I had a final version where all five calculations were working the way they should.

This may seem like overkill, especially if the calculations don't build on one another, but I can't tell you how many times it's saved me. I'll be working on calculation D and suddenly realize that I incorrectly manipulated my data at some point and that I can't undo the change. Instead of having to start over from scratch I can just go back to the point where things were working properly and move forward from there.

(And I know some really sharp readers out there are thinking, oh, but that could be fixed by doing this or that or the other thing. Trust me. While that may be possible, sometimes the easier answer is to just go back to when things were working properly instead.)

Obviously the way you're working with your data will impact the need to save versions or when you save them or how you save them. With R it was a matter of saving versions of the scripts I'd written rather than the data files.

Also, while we're here, let me give you a little tip for naming files that have multiple versions. My preference is to name them with the date in YYYYMMDD format. That way they always sort in the order they were created. So Analysis File 20190101 would be the January 1st 2019 version of Analysis File. If I have two files for a date then I follow that with v1, v2, etc.

Also, if you have an ongoing analysis that you

later change, keep copies of the final draft for all prior versions. Even though you think you may never use the prior version again, there's always that chance that you'll need to go back to see how things differed in the past. (I come from a regulatory background where it's often important how things were being done on a specific date. It's not enough to say, "But this is how we do it now." Plus, there's always that chance that the new way of doing things will turn out to be wrong and you'll want to quickly and easily go back to the old way of doing things.)

TEST YOUR RESULTS

This can be easier said than done, especially with really complex calculations or large data sets, but you should always attempt to test any analysis results that you generate.

For example, if you're calculating sales tax on all of the orders in your database and using a formula to do so, you should actually look at a few specific entries and make sure that the result makes sense. And not just on the first entry.

In Excel, for example, I've seen issues when a formula referenced a value in a fixed cell, but where the formula didn't lock the cell reference (using $ signs), which meant that all entries after the first one were calculating incorrectly. (Something that's usually pretty obvious, but not always.)

I've also seen situations where an edge case wasn't written properly in the formula. (I'm guilty of this one.)

For example, let's say you have different discount percentages that are applied for different order sizes. Spend over $50 and you get 10% off, spend over $100 and you get 15% off. That sort of thing.

In that case you'd want to test order values right at the threshold such as $49.99, $50, and $50.01 and compare that to what's supposed to happen. In this example there should be no discount at $49.99 or $50, but there should be one at $50.01. If it were worded that you get 10% off when you spend $50 or more, then you'd expect to see a discount for a $50 transaction as well.

Those are some of the more obvious issues to test for when you're analyzing data, but another issue to look for is that you have the right data and that it's being used where it should be.

I had a situation once where we were provided the same data in two different formats and didn't realize it. Because of that the data ended up being used twice. This wasn't a math issue. It couldn't be caught by checking a formula. It was an input issue.

The only way to know what had happened was to have enough of an understanding of the data and of the expected outcome of the calculation to see that the result was wrong. This is why involving someone on the business side is so crucial when working with data. They can't tell you, "hey, you used that data set twice" but what they can say is, "no, that doesn't make sense."

So if you ever find yourself in a situation where someone with business knowledge of the situation is saying the result doesn't make sense and the person who did the analysis is saying that the math is correct, check for issues in what data was provided and how it was used. Either it wasn't the right data, it wasn't cleaned up correctly, it was used improperly, or something else that I'm not thinking of now because you can never think of every outcome.

If you're on the business side in that situation, have the guts to speak up and say something doesn't look right. And if you're on the data side, check your ego, listen, and consider that something isn't working even if the issue isn't actually with what you did.

MAKE ASSUMPTIONS VISIBLE

Another principle that goes hand-in-hand with testing your results is making sure that all of the assumptions that went into a calculation are visible so that everyone can see what they were and evaluate their appropriateness.

Let me give you an example that many will be somewhat familiar with.

Let's say you're looking at selling your house and you figure that to sell your house you'll have to pay 6% in realtor fees, 2% for any issues found during the inspection, 1% for cleaning the place up before you list it, $1,000 for moving expenses, and that the house is currently worth $350,000.

Based on all of that information you list the value of your house as $317,500.

The worst way to do this would be to simply list the $317,500 amount. Because two weeks later you may not remember where that number came from. You may remember the $350,000 and the 6% realtor fees, but not understand where the rest of the $11,500 you took off the sale price came from. So then you think you can get more than you can and make a bad decision as a result.

Better to at least have the calculation you performed in the field where you store the value. So something like:

$$=(350000*(1-.06-.02-.01))-1000$$

Then at least you know what value you put on the property and that you made four adjustments to that price. But you'd still have to spend some time thinking about where those four adjustments came from.

The best option is to list out each assumption in its own separate, labeled field that is immediately visible and then to have the final value shown at the end. Doing it that way lets everyone easily see the assumptions that went into the calculation. If six months later you decide the house is worth $375,000 or that your best friend who is a realtor will give you a break and it'll only cost 4% in realtor fees, you can easily see that your current assumptions are incorrect and need to be changed.

Listing each value by itself also makes it easier

to change them. You don't have to mess with the core formula and worry that you've deleted something that shouldn't be deleted. You instead are very clearly changing "sale price", for example.

Let me add here that even as I was writing this there was a part of me that was remembering some of the lovely people I've worked with in the past who would spend hours wanting to debate every single assumption. And that there were times when I knew it would save me a headache and hours of my life to just give them the final value without showing them all of the assumptions that went into calculating it.

So I understand that there can be a temptation to not show all of your assumptions just because you don't want to go through the hassle of having someone question each and every one. But I will reiterate that it is a best practice to make all assumptions visible no matter how painful the resulting conversation might be.

KNOW THE LIMITATIONS OF YOUR PROGRAMS

When working with data, you also need to have a very good understanding of the tools you're using to analyze your data and what the potential limitations of those tools might be.

Let me give you an example of what I mean by this.

I had a project where I was working with data that was stored in a SQL database. I exported that data into a .csv file that I then looked at in Excel.

Part of the data set included the year of incorporation for various companies, some of which had been in existence since the 1800s. That field exported as a date, which it technically was. When I reviewed the data in Excel, Excel had converted those incorporation dates into 1900s dates. So January 1, 1823 became January 1, 1923.

Turns out Excel has limitations in how it works with dates. By default it doesn't work with dates prior to January 1, 1900. (I wrote a whole chapter on this in *50 More Excel Functions* or you can look it up in Excel's help text if you are curious and weren't already aware of the issue.)

At the time I exported that data I wasn't aware of this. Fortunately, I caught the issue and it wasn't material to any calculation we were doing. But knowing something like that could have been very important.

Another example from Excel is in using the count functions. COUNT counts the number of cells in a range that contain numbers, COUNTA counts the number of cells in a range that are not empty. Depending on the entries you're trying to count, this can be a very important difference. If you have mixed numeric and text values and you use COUNT, you will fail to count any cell with a text value.

This is why you should always, always gut check your analysis and have that gut check done by someone with a business-level knowledge of the data and what the results should look like. Data does not exist in a vacuum of pure analysis. It exists within the context of your business and needs to make sense within that context.

KNOW THE DIFFERENCE BETWEEN A REPORT AND RAW DATA

This one goes to how you provide your analysis results to other users.

There is a difference between providing someone with data and providing them with a report about that data, and it's important to understand when to create a data extract and when to create a report.

Let me give you an example from my publishing.

I take sales data from about a dozen sources and combine that data in an Access database that lets me track my revenue and profit by author, series, and title.

That means that what I need from the places that sell my books is a data extract. I don't want a report on my sales or my ad spend, I want a data file that I can incorporate into my systems.

But some of the places where I sell my books insist on providing a report. The worst do so in a PDF format. To use that information in my database I have to manually input it which is time consuming and creates the potential for user error since I'm retyping the data into my system.

Others manage to do so even though they're providing an Excel or a .csv file. For years Amazon provided a monthly summary of sales in an Excel file that was a report not a data extract. They had on one Excel worksheet separate header rows and summaries for each and every store where Amazon sells books, each store separated by a few blank lines. So there'd be a section for amazon.com with an unknown number of rows that varied based on sales for the month followed by a few blank rows and then an unknown number of rows for amazon.co.uk and so on and so on through every single store, even ones with no sales.

Finally, Amazon transformed that report into a data extract. The way they did this was by adding a column that indicated what store the transaction occurred in and then listing all data for all stores in one continuous series of rows and with no summary rows at the end. The old report took me ten minutes to work with, the data extract takes me one. (A slight exaggeration, but not by much.)

To provide the correct type of output to your users, you need to understand how the information you're providing will be used. (To make it trickier there are going to be some users who use what you give them one way and some who use what you give them in a completely different way. I'm sure some of the non-technical writers I know were very unhappy to lose their summaries when Amazon changed how it delivered that information. That's why I wrote them a book on how to use pivot tables.)

It may in fact be necessary to provide both a report of final numbers and a data extract. The first worksheet in an Excel file can be the summary and the next can be the detailed data.

And for the record, this is how I define the two types:

A report is a final product for someone to skim through and say, "Ah, okay, that's where we are."

A data extract is a simple listing of the data with one single row for the header and one single row for each data entry. There should be no extra row for units under the header, there should be no summary information at the top about what's contained in the data, there should be no subtotals or sections within the data, there should only be one row per entry even if that means duplicating data, and there should be no summary below or in the last column of data.

VIEW YOUR DATA FROM
A BUSINESS PERSPECTIVE

Okay, so that's pretty much it, but I just want to reiterate this one more time:

It's very important that someone with a knowledge of your specific business analyze your data and the results of any analysis that you perform on that data.

Someone needs to be able to do a gut check of the results and say if they makes sense.

For example, because I seem to love examples, let's say that you have a program that calculates individual net worth. This program adds up all assets and subtracts all liabilities and generates a final number. And let's say that number is $300 million.

Is that number right?

I don't know.

Maybe. Maybe not.

If all you know is the data side of things then all you can say is that the calculation was right given the inputs. Someone told you that the assets involved added up to $500 million and that there were $200 million in liabilities and when you subtracted liabilities from assets you got a result of $300 million.

Okay.

But does that number make sense given the user who input the data?

Maybe there was user error and the person inputting the values didn't see that they were supposed to input values such as $30,000 as 30. Maybe they input that as 3000000 instead because they included the cents as well.

From a pure data perspective no one can say whether that $300 million result makes sense. Only someone with knowledge of the actual situation and circumstances can.

Me, someone tells me my net worth is $300 million, I'm going to burst out laughing. But some people I've known over the years might say, "Really, is it that low?"

So always have someone involved in handling and analyzing your data who can do that "does this make sense" check of the data.

And force them to do it.

I mentioned earlier that there was a situation where we were given the same dataset twice in two different forms and so it was used twice. In that situation had the subject matter expert stopped to really look at the results of the analysis they would have seen that something was off. The numbers did not make sense from a business perspective. They made sense from an A times B times C perspective, but not from a gut check perspective.

So build it into your process to check the data that comes in and the outcome of your analysis from both a technical perspective and a business perspective. It will save you a lot of grief if you do so.

BUILD IN CHECKS AND BALANCES WHERE YOU CAN

That last discussion made me realize that there is one more data principle to tell you about. Whenever possible, build in checks and balances to make sure that you have manipulated your data appropriately and that all imports, etc. are working properly.

Let me give you an example.

A while back Amazon moved all of the paperbacks that were being sold through their affiliate CreateSpace over to their KDP platform. Somehow during that process they dropped a few of my books. They were listed in my old account but not listed in my new one.

It would've been a relatively simple process to run a comparison after the import that showed number of titles on CreateSpace for that account and to compare that to the number of titles now listed on the new account with additions and deductions for titles that couldn't come over or ones that had already been there.

Clearly they didn't do that. (And I'd say they clearly didn't anticipate that someone might have more than one version of a book published. In my case I had color and black and white versions. Others had large print versions.) I had to write them and inform them of the issue.

They also at one point after that migration dropped all print sales of my top-selling title from my account. It was half of my sales at the time, so a pretty obvious difference. Once more, I had to identify it for them.

But there are any number of ways to build in a check to make sure something like that doesn't happen. You can look at trends in numbers over time. So do today's numbers match with yesterday's within some sort of acceptable threshold? What about this month to last month? Or this month this year to the same month last year?

Another way to do it is to look at the output from two systems or two different reports that use the same data but in different ways.

So when I import a data file into my Access database I make sure to check the total dollar value in the file that I'm going to import against the value for that platform for that month after I import the data since they should match.

I also keep a listing of total revenue by platform that I compare to my final report which shows total sales across all platforms and formats. Those should always match as well, but sometimes I need to add a product identifier for a book on a specific platform before my final sales report will pull in those sales.

Whatever data import you do or data analysis, always try to build in some sort of checks and balances to make sure you're not losing data along the way and that your calculations are reasonable.

ACTIVITY-BASED COSTING

Okay. Almost done.

I told you I was going to discuss activity-based costing, though, so let's do that.

Why am I doing this? One, because I find it such a useful concept and, two, because it's a good example of how even the smallest business can use data analysis to make better decisions.

So the basic concept behind activity-based costing is that you find a way to assign your costs to your products, even those costs that don't seem easy to assign directly.

Let me use my publishing business as an example of how to do this.

Let's say that I publish two books in a year and I have revenue from each of those books. I also have advertising, cover, and editing expenses specific to each one. Those are my direct costs.

Book A earns $5,000 and has direct costs of $3,000 so generates a $2,000 profit.

Book B earns $10,000 and has direct costs of $4,000 so generates a $6,000 profit.

But let's say I also spend $5,000 a year on things like my website, phone, internet, etc. They're business expenses, but they don't tie directly to either book. These are my indirect expenses.

Without activity-based costing, I'd probably either not assign the indirect expenses to either book, in which case I'd say I had an overall profit of $3,000 which is the per-book profits of $2,000 and $6,000 minus the $5,000 in other costs.

Or I'd divvy it up evenly between the two titles. So Book A would get an additional $2,500 in expenses and now show a loss of $500 and Book B would get an additional $2,500 in expenses and show a profit but of only $3,500.

That might not be a fair way to divide those costs, though.

What if Book A took me only twenty hours to write and Book B took me two hundred hours? If one book took ten times the amount of effort to produce, shouldn't that somehow be factored into how I assign those indirect costs?

I like to think so.

To do this we can use the hours spent writing to create a ratio that lets us assign the indirect expenses.

Let's walk through it.

In this example, the total number of writing hours for the year is 220 hours (20 plus 200).

Book A was 9% of that time (20/220) and Book B was 91% of that time (200/220). (It's actually 9.09% and 90.9% but we're just going to go with 9% and 91% for the rest of this example.)

If I then assign the $5,000 in indirect expenses using those percentages, I have indirect expenses for Book A of $450 and a revised profit of approximately $1,550. And I have indirect expenses for Book B of $4,550 and a revised profit of $1,450.

Using activity-based costing Book A is now more profitable than Book B.

See how powerful that can be?

Now, one of the tricks with activity-based costing is figuring out which metric to use to assign indirect costs. I could have used word count instead of hours spent writing if that made more sense. Or number of titles if I were trying to assign costs across pen names. Or number of fan mails maybe if I had an admin who handled things like that. You have to think about what makes the best sense given your circumstances. And maybe even use different metrics for different indirect expenses.

It's kind of an art more than a science.

Also, I hope you see how we couldn't have

done the above calculation if we hadn't been tracking the right data. In this case, time spent writing.

CONCLUSION

Alright. So those are the principles or guidelines I use when dealing with data. As you can see they range across everything from what data to collect to how to collect it to how to store it to how to use it all the way to how to analyze it.

I hope you've realized after reading this book that the first step in data analysis is having the right data. That means understanding your business, knowing what information is available to collect, making the right decisions about which data you need or want, collecting it in a way that allows for analysis and comparison, and then performing that analysis in a valid, replicable manner that leaves the source data untouched.

If you can actually do all of that you will have an advantage over your business competition. Because you will be able to see things that they cannot.

It's not necessarily hard to do any of this, but it does require forethought to set things up properly and discipline to implement it well.

Obviously, big data takes what I've shown you here and goes far, far beyond it. And there are certainly highly sophisticated techniques for fixing data after the fact if you get it wrong. But for most businesses you don't need to go there. You can make significant improvements with just some small tweaks and by tracking the right information in the first place.

And don't despair if you haven't been tracking what you need to track, just fix it. Be better tomorrow than you were today.

Hope this helped. If you have any questions, feel free to reach out. I won't necessarily respond within a day, but I will respond.

ABOUT THE AUTHOR

M.L. Humphrey is a former stockbroker with a degree in Economics from Stanford and an MBA from Wharton who has spent over twenty years as a securities regulator, consultant to Fortune 500 financial institutions, and small business owner.

You can reach M.L. at mlhumphreywriter@gmail.com or at mlhumphrey.com.